MASTERING THE CONSTRUCTION BUSINESS

Steven Smith

Wisdom Publishers

Copyright © 2023 Wisdom Publishers

All rights reserved

The characters and events portrayed in this book are fictitious. Any similarity to real persons, living or dead, is coincidental and not intended by the author.

No part of this book may be reproduced, or stored in a retrieval system, or transmitted in any form or by any means, electronic, mechanical, photocopying, recording, or otherwise, without express written permission of the publisher.

ISBN: 9798399811819
Imprint: Independently published

Cover design by: Art Painter
Library of Congress Control Number: 2018675309
Printed in the United States of America

To all the ambitious dreamers, courageous builders, and relentless innovators in the construction industry,

This book is dedicated to you. You, who strive to transform mere blueprints into awe-inspiring structures that shape the world we live in. You, who tirelessly push the boundaries of what's possible and create remarkable legacies for generations to come.

Your unwavering passion, unwavering commitment, and unwavering determination inspire us all. You face countless challenges with resilience, turning obstacles into opportunities. It is your unwavering spirit that drives progress and propels the construction industry forward.

This dedication is a testament to your unwavering dedication, unwavering expertise, and unwavering pursuit of excellence. May this book serve as a guiding light on your journey, offering insights, knowledge, and strategies to conquer the ever-changing landscape of the construction business.

As you master the intricacies of the industry, may you continue to build with integrity, lead with purpose, and leave a lasting impact on the world around you. Your unwavering dedication fuels the transformation of cities, the realization of visions, and the creation of remarkable spaces that shape our lives.

Here's to the dreamers who turn blueprints into reality, the visionaries who shape skylines, and the builders who construct a brighter future. This book is dedicated to you—the true masters of the construction business.

With deepest admiration and respect,

[Steven Smith]

CONTENTS

Title Page
Copyright
Dedication
Introduction
Part I: Understanding the Construction Business Landscape — 1
Chapter 1: Overview of the Construction Industry — 2
Chapter 2: Legal and Regulatory Framework — 6
Chapter 3: Market Analysis and Business Planning — 11
Part II: Essential Business Functions in Construction — 16
Chapter 4: Financial Management — 17
Chapter 5: Project Management — 22
Chapter 6: Risk Management — 28
Chapter 7: Marketing and Sales — 33
Chapter 8: Human Resources Management — 39
Part III: Operational Excellence in Construction Business — 45
Chapter 9: Project Estimation and Cost Control — 46
Chapter 10: Quality Management — 54
Chapter 11: Technology and Innovation — 62
Chapter 12: Sustainability and Green Building — 69
Part IV: Navigating Challenges and Ensuring Success — 73
Chapter 13: Project Delivery Methods — 74

Chapter 14: Ethics and Professionalism   77
Chapter 15: Succession Planning and Business Continuity   80
About The Author   87
Books By This Author   89

INTRODUCTION

The construction industry, with its vast scope and transformative impact on society, stands as a dynamic and demanding domain that requires a meticulous approach to management and a keen understanding of its intricacies. As a professor deeply entrenched in the field of construction management, it is my privilege to present this comprehensive guide, an invaluable resource for those seeking to navigate the ever-evolving landscape of the construction business.

In today's fast-paced world, success in the construction industry extends far beyond technical know-how and construction methodologies. It hinges on a holistic comprehension of business principles, strategic planning, effective project execution, robust financial management, risk mitigation, marketing prowess, and the ability to foster a resilient and adaptable organizational culture. This book endeavors to provide you with a multidimensional perspective on construction management, equipping you with the tools and knowledge necessary to excel in this challenging arena.

Throughout my career, I have witnessed firsthand the triumphs and tribulations faced by construction professionals and businesses. I have seen projects that soared to great heights, becoming architectural marvels that shaped skylines and transformed communities. Conversely, I have also witnessed the unfortunate consequences of inadequate planning, flawed execution, and the lack of a solid business foundation leading to setbacks and even failure. It is from this wealth of experience that this book takes shape, aiming to guide you towards success while helping you avoid the common pitfalls that can hinder progress in the construction business.

Our journey together begins by laying a strong foundation in understanding the construction industry's landscape and its historical evolution. We explore the key stakeholders involved, the economic forces at play, and the trends that are shaping the industry's trajectory. By comprehending the broader context, you will be better positioned to identify opportunities and anticipate challenges that lie ahead.

To navigate the complexities of the construction business, we delve into the legal and regulatory framework that underpins its operations. Understanding the intricacies of contracts, permits, and compliance requirements will empower you to navigate these legal realms with confidence and avoid potential legal pitfalls that can impact your projects and business outcomes.

A crucial aspect of success in the construction industry lies in effectively analyzing the market landscape and formulating a robust business plan. We delve into the methodologies and tools of market research, enabling you to identify target markets, understand customer needs, and craft a comprehensive strategy that differentiates your business from the competition. By developing a solid business plan, you will have a blueprint for success that aligns your goals, resources, and capabilities.

Financial management is a cornerstone of any successful business, and the construction industry is no exception. In this book, we explore the principles of financial planning, budgeting, cash flow management, cost estimation, and profitability analysis. Understanding these essential concepts and applying them to your construction business will help you make informed financial decisions, allocate resources efficiently, and achieve sustainable growth.

Effective project management is the key to delivering successful construction projects on time, within budget, and to the desired quality standards. We dive deep into the intricacies of project initiation, planning, execution, monitoring, control, and closure. By adopting best practices in project management, you will be equipped to lead teams, manage resources effectively, mitigate risks, and overcome the myriad challenges encountered in the construction journey.

Risk management, an inherent part of the construction business, holds the key to safeguarding your projects and your organization's reputation. We explore the identification, assessment, and mitigation of risks, along with crisis management strategies and business continuity planning. By proactively addressing risks and developing resilient strategies, you can minimize potential disruptions and ensure the smooth progress of your construction projects.

No business can thrive without a strong focus on marketing and sales. In this book, we delve into the intricacies of branding, positioning, lead generation, customer relationship management, bidding, and proposal preparation. By mastering these essential marketing techniques, you can enhance your business's visibility, attract the right clients, and build lasting partnerships that contribute to your long-term success.

The construction industry is driven by human capital, making effective human resource management a critical factor in your business's success. We explore workforce planning, recruitment strategies, training and development, performance management, leadership, and diversity and inclusion. By nurturing a skilled and motivated workforce, you can create a culture of excellence, attract and retain top talent, and cultivate a high-performing organization.

Operational excellence lies at the heart of a successful

construction business. We delve into topics such as project estimation, cost control, quality management, technological innovation, and sustainability. By embracing efficient operational practices and leveraging emerging technologies, you can optimize your construction processes, enhance productivity, and deliver superior results to your clients.

Throughout this book, we draw upon real-world case studies, practical examples, and lessons learned from successful construction businesses. By examining both triumphs and challenges, we aim to provide you with a well-rounded perspective on the construction industry and the practical tools necessary to overcome obstacles and achieve your business objectives.

The construction business is an exciting and rewarding field, filled with opportunities for growth, innovation, and making a lasting impact on the built environment. By embarking on this educational journey together, we will explore the rubrics of the construction business, unlocking the strategies and insights that will empower you to excel and thrive in this ever-evolving industry.

Now, let us begin this transformative exploration, charting a path towards mastery in the construction business, and embark on a journey that will shape the success of your endeavors.

PART I: UNDERSTANDING THE CONSTRUCTION BUSINESS LANDSCAPE

CHAPTER 1: OVERVIEW OF THE CONSTRUCTION INDUSTRY

The construction industry stands as a vital pillar of economic development, shaping the physical infrastructure that supports societies around the world. This chapter provides a comprehensive overview of the construction industry, examining its evolution, growth, key players, stakeholders, economic impact, market trends, as well as the challenges and opportunities it presents.

1.1 Evolution and Growth of the Construction Industry

The construction industry has evolved significantly over time, reflecting changes in technology, societal needs, and economic landscapes. From its earliest forms of manual labor and basic tools to the sophisticated machinery and advanced construction techniques of today, the industry has undergone a transformative journey.

Historically, construction was primarily driven by basic human needs, such as shelter and infrastructure. As civilizations advanced, construction techniques became more complex, enabling the creation of monumental structures and architectural wonders. The Industrial Revolution marked a significant turning point, introducing mechanization, mass production, and the use of new materials, propelling the industry forward.

In recent decades, the construction industry has witnessed accelerated growth due to rapid urbanization, population growth, and infrastructure demands. Advancements in technology, such as Building Information Modeling (BIM), prefabrication, and sustainable construction practices, have revolutionized the industry, enabling greater efficiency, cost-effectiveness, and environmental responsibility.

1.2 Key Players and Stakeholders in Construction

The construction industry encompasses a diverse range of professionals, organizations, and stakeholders. Key players include construction companies, contractors, architects, engineers, project managers, suppliers, subcontractors, regulatory bodies, and clients. Each entity plays a crucial role in the construction process, contributing their expertise and resources to bring projects to fruition.

Understanding the roles and responsibilities of these stakeholders is vital for effective collaboration, coordination, and successful project outcomes. Construction companies serve as the driving force behind project execution, responsible for planning, organizing, and managing resources to deliver construction projects. Architects and engineers provide design expertise, ensuring structural integrity, functionality, and aesthetics. Project managers oversee the entire project lifecycle, coordinating various stakeholders and ensuring project goals are achieved.

Suppliers and subcontractors play a vital role in the supply chain, providing essential materials, equipment, and specialized services. Regulatory bodies enforce standards and regulations, ensuring compliance with safety, environmental, and building codes. Clients, whether private individuals, businesses, or government entities, drive the demand for construction projects, setting project objectives, and providing financial resources.

1.3 Economic Impact and Market Trends

The construction industry significantly contributes to national economies, creating employment opportunities, driving investment, and stimulating economic growth. Construction projects have a multiplier effect, generating demand for materials, labor, and services across various sectors, thereby enhancing economic activity.

Market trends in the construction industry are influenced by various factors. Economic conditions, such as GDP growth, interest rates, and government policies, impact construction activity. Demographic shifts, urbanization, and population dynamics shape the demand for residential, commercial, and infrastructure projects. Technological advancements, sustainability concerns, and changing consumer preferences drive innovation and influence market trends.

Understanding these economic and market dynamics is crucial for construction businesses to make informed decisions, identify emerging opportunities, and adapt to changing conditions. By staying abreast of market trends, construction professionals can position themselves strategically and align their offerings with market demands.

1.4 Challenges and Opportunities in Construction Business

The construction business is not without its challenges. Complex regulatory frameworks, stringent safety requirements, tight budgets, schedule constraints, and labor shortages pose significant hurdles to successful project delivery. Uncertainty in economic conditions, market volatility, and intense competition further add to the challenges faced by construction businesses.

However, within these challenges lie ample opportunities for growth and innovation. Technological advancements, such as digitalization, robotics, and artificial intelligence, offer new avenues for improving productivity, efficiency, and safety. Sustainable construction practices and green building initiatives address environmental concerns while providing a competitive edge. Collaborative project delivery methods, such as Building Information Modeling (BIM) and Integrated Project Delivery (IPD), foster enhanced communication, coordination, and collaboration among stakeholders.

The growing demand for infrastructure development, urban renewal, and energy-efficient buildings present significant opportunities for construction businesses to expand their portfolios and meet societal needs.

By proactively addressing challenges, embracing innovation, and capitalizing on emerging opportunities, construction businesses can thrive in a rapidly evolving industry.

CHAPTER 2: LEGAL AND REGULATORY FRAMEWORK

2.1 Laws and Regulations in Construction Business

The construction industry operates within a complex web of legal and regulatory frameworks that govern various aspects of project execution. Understanding the laws and regulations applicable to the construction business is essential for compliance, risk management, and ensuring the safety and well-being of all stakeholders involved.

Legal requirements encompass a wide range of areas, including labor laws, contract laws, tort laws, property laws, environmental laws, and health and safety regulations. These laws aim to protect the rights of workers, promote fair business practices, ensure the quality and safety of construction projects, and mitigate potential risks and liabilities.

Construction businesses must familiarize themselves with the specific laws and regulations applicable to their jurisdiction. This includes understanding federal, state, and local laws, as well as industry-specific regulations governing construction activities. By staying up to date with legal requirements, construction professionals can mitigate legal risks, avoid penalties, and maintain the integrity of their operations.

2.2 Licensing and Permitting Requirements

Licensing and permitting requirements play a vital role in the construction industry, ensuring that construction activities are carried out by qualified professionals and in compliance with established standards. Licenses are typically required for contractors, architects, engineers, and other professionals involved in construction projects.

Licensing requirements vary by jurisdiction and may include educational qualifications, professional experience, and successful completion of licensing examinations. By obtaining the necessary licenses, construction professionals demonstrate their competence and commitment to adhering to industry standards and regulations.

Permitting requirements, on the other hand, pertain to obtaining approvals from regulatory authorities before commencing construction activities. Permits may be required for various aspects of the project, such as building permits, electrical permits, plumbing permits, and environmental permits. Compliance with permitting requirements ensures that construction projects meet safety, environmental, and building code standards.

2.3 Compliance with Building Codes and Standards

Building codes and standards are a critical component of the legal and regulatory framework in the construction industry. They provide guidelines and requirements for the design, construction, and maintenance of buildings to ensure safety, structural integrity, and occupant well-being.

Building codes are generally adopted at the national, state, or local level and cover various aspects of construction, including structural design, fire safety, electrical systems, plumbing, accessibility, and energy efficiency. Compliance with building codes is mandatory and enforced by regulatory authorities.

In addition to building codes, industry standards and guidelines set forth best practices and technical requirements for specific construction activities. These standards, developed by professional organizations and regulatory bodies, cover areas such as construction materials, construction methods, safety practices, and quality control. Adhering to industry standards enhances the quality of construction projects and helps mitigate risks.

2.4 Environmental and Safety Regulations

Environmental regulations play a crucial role in the construction industry, aiming to protect the environment, conserve natural resources, and promote sustainable practices. Construction projects have the potential to impact air quality, water resources, ecosystems, and the overall environmental balance. Compliance with environmental regulations is essential for minimizing these impacts and ensuring environmentally responsible construction practices.

Environmental regulations may include requirements for managing construction waste, controlling erosion and sedimentation, mitigating noise and air pollution, protecting endangered species, and promoting energy efficiency and renewable energy use. Construction businesses must integrate environmental considerations into their project planning, implementation, and monitoring processes to achieve compliance and contribute to sustainability efforts.

Safety regulations are of paramount importance in the

construction industry due to the inherent risks associated with construction activities. Construction sites are dynamic and complex, involving heavy machinery, hazardous materials, elevated work areas, and various potential safety hazards.

Occupational safety and health regulations establish requirements to protect workers from injuries, illnesses, and fatalities in the construction industry. These regulations encompass areas such as hazard identification and assessment, personal protective equipment (PPE), fall protection, electrical safety, machine guarding, and emergency preparedness.

Compliance with safety regulations involves implementing effective safety management systems, conducting regular safety inspections, providing proper training to workers, and promoting a culture of safety within the organization. Prioritizing safety not only protects workers' lives and well-being but also mitigates legal risks, improves productivity, and enhances the reputation of construction businesses.

2.5 Contracts and Legal Considerations

Contracts form the foundation of business relationships in the construction industry, outlining the rights, responsibilities, and obligations of parties involved. Understanding contract law and implementing sound contracting practices is essential for protecting the interests of construction businesses, managing risks, and avoiding disputes.

Construction contracts typically cover various aspects, such as project scope, specifications, schedule, payment terms, change orders, dispute resolution mechanisms, and liability provisions. These contracts may involve multiple parties, including owners, contractors, subcontractors, suppliers, and design professionals.

Legal considerations in contract management include ensuring contract clarity and completeness, assessing contractual risks, documenting contract changes, and implementing effective contract administration processes. Proper contract management facilitates project coordination, minimizes disputes, and provides a solid legal foundation for successful project execution.

By navigating the legal and regulatory landscape of the construction industry effectively, construction professionals can ensure compliance, protect the interests of all stakeholders, and create a framework for safe and successful project outcomes.

CHAPTER 3: MARKET ANALYSIS AND BUSINESS PLANNING

In the construction industry, market analysis and business planning are crucial components for achieving long-term success and sustainable growth. This chapter explores the key considerations involved in conducting a comprehensive market analysis and developing a strategic business plan tailored to the construction business.

3.1 Identifying Target Markets and Customers

To effectively position a construction business in the marketplace, it is essential to identify target markets and customers. This involves understanding the specific sectors, geographical locations, and demographics that present the greatest opportunities for growth and demand.

Target markets in the construction industry can include residential, commercial, industrial, institutional, or public infrastructure sectors. Each market segment has unique characteristics, challenges, and requirements. For instance, the residential market may focus on single-family homes, multi-unit residential buildings, or remodeling and renovation projects. The commercial market may encompass office buildings, retail spaces, or hospitality establishments.

Identifying the target customer within each market segment is equally important. This could be a private homeowner, real estate developer, government agency, or commercial enterprise. Understanding the needs, preferences, and pain points of these customers is crucial for tailoring products and services to meet their specific requirements.

3.2 Market Research and Analysis

Market research and analysis serve as the foundation for informed decision-making and strategic planning. It involves gathering and analyzing relevant data and information about the construction industry, market trends, customer preferences, and competitive landscape.

Primary research methods, such as surveys, interviews, and focus groups, can provide valuable insights directly from customers, industry experts, and key stakeholders. Secondary research, including industry reports, market studies, and data analysis, helps to identify industry trends, market size, growth potential, and competitor analysis.

Case Study: A consulting firm looking to expand its services into the commercial sector conducts market research. Through interviews with commercial property developers and analysis of industry reports, they identify a growing demand for sustainable construction practices and LEED certification. Armed with this information, they develop a strategic plan to offer specialized consulting services in sustainable construction practices, positioning themselves as experts in the market.

3.3 Developing a Business Plan

A comprehensive business plan acts as a roadmap for achieving business objectives and guides decision-making processes. It outlines the mission, vision, values, target markets, competitive advantages, and operational strategies of the construction business.

The business plan should include sections on marketing and sales strategies, organizational structure, financial projections, risk management, and implementation timelines. It provides a clear framework for resource allocation, budgeting, and performance evaluation.

Case Study: A construction company specializing in commercial building projects develops a business plan to expand its operations into the healthcare sector. Their plan includes a detailed market analysis of the healthcare industry, a targeted marketing strategy to reach healthcare providers, and a comprehensive risk management plan to address the unique challenges of building healthcare facilities. The business plan provides a roadmap for successfully entering this new market segment.

3.4 Defining Unique Selling Proposition (USP)

In a competitive marketplace, establishing a Unique Selling Proposition (USP) sets a construction business apart from its competitors. The USP identifies and communicates the distinctive features, advantages, and benefits that differentiate the business from others in the market.

The USP could be based on various factors, such as specialized expertise, innovative construction methods, sustainability practices, exceptional customer service, or a strong track record of successful projects. It should resonate with the target customers and align with their needs and priorities.

Case Study: A construction company differentiates itself by specializing in complex renovation projects for historic buildings. They showcase their expertise in preserving historical architectural features, incorporating modern amenities, and ensuring compliance with preservation guidelines. This unique selling proposition attracts clients who value the restoration of historical structures while requiring functional and contemporary spaces.

3.5 Setting Goals and Objectives for Business Success

Setting clear goals and objectives is vital for directing the efforts of a construction business and measuring its progress towards success. Goals should be specific, measurable, achievable, relevant, and time-bound (SMART), providing a framework for performance evaluation and continuous improvement.

Business goals can include financial targets, market share expansion, customer satisfaction levels, employee training and development, safety performance, or sustainability objectives. Each goal should be accompanied by actionable strategies, key performance indicators (KPIs), and milestones to track progress.

Case Study: A construction management firm sets a goal to reduce workplace accidents by 30% within the next year. They implement a comprehensive safety training program, enforce strict safety protocols, and conduct regular safety audits. Through continuous monitoring and improvement, they successfully achieve their goal, resulting in a safer working environment and reduced project delays.

By conducting thorough market analysis and developing a well-structured business plan, construction professionals can position their businesses for success in the ever-evolving construction industry. These strategic processes enable businesses to identify lucrative market opportunities, differentiate themselves from competitors, and set realistic goals for growth and sustainability.

PART II: ESSENTIAL BUSINESS FUNCTIONS IN CONSTRUCTION

CHAPTER 4: FINANCIAL MANAGEMENT

Financial management is a critical aspect of running a successful construction business. This chapter explores the key elements of financial management specific to the construction industry, including financial planning, cash flow management, cost estimation, financial statements analysis, and financing options.

4.1 Financial Planning and Budgeting

Effective financial planning and budgeting lay the groundwork for sound financial management in construction businesses. It involves forecasting revenue, estimating expenses, and allocating resources to achieve financial objectives.

Financial planning begins with setting financial goals, such as revenue targets, profitability ratios, and return on investment (ROI) benchmarks. These goals should be aligned with the overall business strategy and take into account market conditions, project pipelines, and industry trends.

Budgeting involves translating the financial plan into a detailed roadmap for resource allocation. It entails estimating costs for labor, materials, equipment, subcontractors, overhead expenses, and contingencies. By developing accurate and comprehensive budgets, construction businesses can effectively control costs, monitor performance, and make informed financial decisions.

Case Study: A construction company specializing in residential projects develops a financial plan and budget for the upcoming fiscal year. Through thorough analysis of historical project data, market trends, and anticipated expenses, they forecast a revenue increase of 15%. They allocate resources accordingly, including funds for hiring additional staff, upgrading equipment, and implementing marketing initiatives to support their growth goals.

4.2 Cash Flow Management and Profitability Analysis

Cash flow management is crucial for the financial health and stability of construction businesses. It involves monitoring and optimizing the inflow and outflow of cash to ensure sufficient liquidity for ongoing operations, timely payment of expenses, and investment in future growth.

Construction businesses often face unique cash flow challenges due to the nature of the industry, such as project-based billing, delayed payments, and fluctuations in project timelines. Effective cash flow management strategies include accurate invoicing, diligent collection efforts, negotiating favorable payment terms with clients, and maintaining strong relationships with suppliers and subcontractors.

Profitability analysis is another critical component of financial management. It involves assessing the profitability of individual projects, project types, or market segments. By analyzing costs, revenues, and profit margins, construction businesses can identify areas for improvement, evaluate the success of pricing strategies, and make informed decisions regarding project selection and resource allocation.

Case Study: A construction contractor conducts a profitability analysis of their past projects. By comparing project costs, revenues, and profit margins, they identify that their commercial building projects consistently yield higher profitability compared to their residential projects. Armed with this knowledge, they shift their focus towards securing more commercial contracts and adjust their business strategy accordingly to maximize profitability.

4.3 Cost Estimation and Pricing Strategies

Accurate cost estimation is crucial for construction businesses to prepare competitive bids, negotiate contracts, and ensure profitability. It involves estimating the costs of labor, materials, equipment, subcontractors, overhead expenses, and contingencies associated with a construction project.

Construction businesses employ various cost estimation techniques, including unit cost estimating, parametric estimating, and detailed quantity take-offs. Leveraging historical project data, industry benchmarks, and expert judgment, cost estimators can develop reliable cost estimates that account for project-specific variables and risks.

Pricing strategies in the construction industry require a careful balance between competitiveness and profitability. Construction businesses must consider market rates, project complexity, profit margins, and client expectations when determining their pricing. Effective pricing strategies involve accurately estimating costs, understanding the value proposition offered to clients, and considering long-term business objectives.

Case Study: A construction subcontractor specializes in electrical installations. When preparing a bid for a new project, they conduct a detailed cost estimation that factors in the project scope, specifications, materials, labor, and potential risks. By considering their costs, competitive pricing, and market demand, they develop a pricing strategy that allows them to secure the project while maintaining a healthy profit margin.

4.4 Financial Statements and Analysis

Financial statements provide a snapshot of a construction business's financial performance and position. They include the balance sheet, income statement, and cash flow statement, each serving a unique purpose in assessing the financial health of the business.

The balance sheet provides an overview of the business's assets, liabilities, and equity at a specific point in time, enabling stakeholders to assess solvency, liquidity, and capital structure. The income statement presents revenues, expenses, and net income over a specific period, offering insights into profitability and operational efficiency. The cash flow statement tracks the inflows and outflows of cash, allowing businesses to assess their ability to generate and manage cash effectively.

Financial statement analysis involves examining these statements to evaluate a construction business's performance, profitability, liquidity, and financial stability. Key financial ratios, such as gross profit margin, net profit margin, return on investment, and current ratio, help assess the business's financial health, compare performance over time, and benchmark against industry standards.

Case Study: A construction company reviews its financial statements to assess its financial performance for the previous fiscal year. By analyzing the income statement, they identify a decrease in gross profit margin compared to industry averages. This prompts further investigation into cost overruns and inefficiencies, leading to the implementation of cost control measures and process improvements to enhance profitability.

4.5 Financing Options for Construction Businesses

Construction businesses often require access to various financing options to support their operations, invest in growth, and manage cash flow fluctuations. Understanding the available financing options and selecting the most suitable ones is essential for ensuring financial stability and supporting business objectives.

Common financing options in the construction industry include bank loans, lines of credit, equipment leasing, trade credit, and equity financing. Each option has its advantages and considerations in terms of interest rates, repayment terms, collateral requirements, and impact on the business's financial position.

Case Study: A construction startup seeks financing to purchase new equipment and expand its capacity. After evaluating different financing options, they decide to secure a bank loan with favorable terms. The loan provides the necessary funds to acquire the equipment, support their growth plans, and strengthen their competitive position in the market.

By effectively managing financial aspects such as planning, budgeting, cash flow, cost estimation, financial analysis, and financing options, construction professionals can ensure the financial stability and long-term success of their businesses. These financial management practices are integral to making informed decisions, optimizing profitability, and fostering sustainable growth within the construction industry.

CHAPTER 5: PROJECT MANAGEMENT

Project management plays a pivotal role in the success of construction endeavors. It encompasses the initiation, planning, execution, monitoring, control, and closure of projects to achieve project objectives within defined constraints of time, cost, quality, and scope. This chapter delves into the key aspects of project management in the context of the construction industry, including project initiation and planning, project organization and team building, project scheduling and time management, resource allocation and procurement, and project monitoring, control, and closure.

5.1 Project Initiation and Planning

Successful project initiation and planning set the stage for a well-executed and controlled construction project. During this phase, project objectives, scope, constraints, and stakeholders' requirements are identified and documented. The initiation process involves establishing the project's feasibility, conducting a thorough needs assessment, and defining the project's purpose and goals.

Once the project is initiated, meticulous planning is crucial to ensure a comprehensive roadmap for project execution. Planning activities encompass defining project deliverables, creating a work breakdown structure (WBS), establishing a project schedule, estimating resource requirements, and identifying potential risks and mitigation strategies. Proper planning sets the foundation for efficient resource allocation, cost management, and risk mitigation throughout the project lifecycle.

Case Study: A construction company is awarded a contract to build a commercial complex. During the project initiation phase, they conduct a feasibility study to assess the project's viability, evaluate site conditions, and analyze market demand. Subsequently, they engage in detailed planning activities, including developing a comprehensive WBS, creating a project schedule, and conducting risk assessments. These meticulous planning efforts enable them to streamline project execution and mitigate potential risks effectively.

5.2 Project Organization and Team Building

Effective project organization and team building are critical for cohesive project execution. Construction projects involve multiple stakeholders, including owners, architects, engineers, contractors, subcontractors, suppliers, and regulatory authorities. Establishing clear lines of communication, roles, and responsibilities is essential to foster collaboration and coordination among project team members.

Project organization involves creating a project team structure, appointing key personnel, and defining their roles and responsibilities. Effective team building entails nurturing a collaborative and supportive project culture, fostering open communication, and promoting teamwork and shared objectives. By developing a high-performing project team, construction businesses can enhance project outcomes and minimize conflicts or delays.

Case Study: A construction project manager assembles a project team for a large-scale infrastructure project. The team comprises experienced professionals from different disciplines, including civil engineering, architecture, and construction management. Through regular team meetings, effective communication channels, and team-building activities, the project manager creates a cohesive team that collaborates seamlessly and delivers successful project outcomes.

5.3 Project Scheduling and Time Management

Efficient project scheduling and time management are crucial for meeting project deadlines, optimizing resource utilization, and ensuring timely project completion. Construction projects involve multiple activities, dependencies, and constraints that require careful planning and scheduling.

Project scheduling involves creating a detailed timeline for project activities, considering task dependencies, resource availability, and project constraints. Construction businesses use various scheduling techniques, such as the critical path method (CPM) and the program evaluation and review technique (PERT), to develop realistic project schedules.

Time management focuses on monitoring and controlling project progress against the established schedule. This involves tracking actual progress, identifying delays or variances, and implementing appropriate corrective measures to mitigate potential schedule slippages. Effective time management ensures that projects stay on track, adhere to deadlines, and maintain customer satisfaction.

Case Study: A construction project team develops a

comprehensive project schedule using the critical path method (CPM). By identifying critical activities and their interdependencies, they prioritize tasks, allocate resources effectively, and establish realistic project milestones. Regular monitoring of project progress enables the team to identify potential delays and take proactive measures to ensure on-time project delivery.

5.4 Resource Allocation and Procurement

Efficient resource allocation and procurement contribute to project success by ensuring the availability of the necessary resources, materials, and equipment at the right time and in the right quantities. Construction projects require careful management of labor, materials, equipment, and subcontractors to maintain productivity and meet project objectives.

Resource allocation involves assessing resource requirements, assigning tasks to the appropriate personnel, and optimizing resource utilization throughout the project lifecycle. It requires considering factors such as labor availability, skill sets, equipment availability, and material availability to avoid delays and inefficiencies.

Procurement management focuses on acquiring materials, equipment, and services from external sources. It entails identifying procurement needs, selecting suppliers, negotiating contracts, and managing supplier relationships. Effective procurement practices help ensure timely delivery of materials, adherence to quality standards, and cost-effective project execution.

Case Study: A construction project manager conducts a thorough analysis of resource requirements for a residential construction project. By considering the project's scope, timelines, and budget, they allocate resources efficiently, including labor, equipment, and materials. Through effective procurement practices, they establish partnerships with reliable suppliers, negotiate favorable contracts, and ensure timely delivery of materials and equipment to support project execution.

5.5 Project Monitoring, Control, and Closure

Project monitoring, control, and closure activities are essential for overseeing project progress, managing changes, and ensuring project closure in a controlled and systematic manner. These activities enable construction businesses to maintain project performance, meet project objectives, and facilitate a smooth transition from project completion to the post-construction phase.

Project monitoring involves tracking and assessing project progress, comparing actual performance against planned targets, and identifying deviations or potential risks. Regular monitoring allows project managers to take corrective actions, make informed decisions, and maintain project alignment with predetermined objectives.

Project control entails implementing measures to mitigate risks, manage changes, and maintain project quality, cost, and schedule adherence. It involves effective communication, change management procedures, quality control processes, and continuous performance evaluations.

Project closure involves formalizing project completion, ensuring client satisfaction, and transitioning the project to the post-construction phase. It includes activities such as final inspections, documentation, project handover, lessons learned, and client feedback collection.

Case Study: A construction project team implements a robust project monitoring and control system for a commercial building project. Through regular progress meetings, frequent site inspections, and continuous communication with stakeholders, they proactively identify potential issues, implement necessary adjustments, and maintain project quality, cost, and schedule adherence. Upon project completion, they conduct a comprehensive project closure process, including documentation, client feedback, and capturing lessons learned for future projects.

By effectively managing project initiation and planning, project organization and team building, project scheduling and time management, resource allocation and procurement, and project monitoring, control, and closure, construction professionals can ensure successful project execution. These project management practices facilitate efficient resource utilization, timely project completion, and customer satisfaction within the dynamic and demanding construction industry.

CHAPTER 6: RISK MANAGEMENT

Risk management is an integral part of construction business operations, aiming to identify, assess, mitigate, and monitor risks that may impact project success, financial stability, and overall business operations. This chapter delves into the key aspects of risk management in the context of the construction industry, including identifying and assessing business risks, implementing insurance and risk transfer strategies, developing risk mitigation plans, establishing crisis management and business continuity protocols, and learning from risk to drive business improvement.

6.1 Identifying and Assessing Business Risks

Identifying and assessing business risks is the foundation of effective risk management. Construction businesses face a wide array of risks that can impact their operations, including financial risks, legal and regulatory risks, safety risks, project-specific risks, market risks, and environmental risks.

Identifying risks involves conducting a comprehensive risk assessment, which includes reviewing historical data, analyzing industry trends, conducting site assessments, and engaging stakeholders to understand potential risks and their potential impact. Risk assessment techniques, such as SWOT analysis (strengths, weaknesses, opportunities, and threats) and risk matrices, assist in prioritizing risks based on their likelihood and potential consequences.

Case Study: A construction company embarks on a new project involving the renovation of an existing structure. During the risk identification and assessment process, they identify potential risks such as unforeseen structural issues, hazardous materials, and neighborhood disruptions. By assessing these risks based on their likelihood and potential impact, the company can proactively plan mitigation measures and allocate appropriate resources.

6.2 Insurance and Risk Transfer Strategies

Insurance and risk transfer strategies are vital components of risk management in the construction industry. They provide financial protection against unforeseen events and help mitigate the potential impact of risks on business operations and project outcomes.

Construction businesses must carefully evaluate their insurance needs and select appropriate coverage to address specific risks. Common insurance types in construction include general liability insurance, professional liability insurance, builder's risk insurance, and worker's compensation insurance. Additionally, risk transfer strategies, such as indemnification clauses in contracts or subcontracting certain project elements to specialized firms, can help transfer risk to other parties.

Case Study: A construction company engages in a major infrastructure project with significant financial and operational risks. To mitigate these risks, they secure comprehensive insurance coverage, including builder's risk insurance to protect against property damage during construction and liability insurance to cover potential accidents or injuries. By implementing risk transfer strategies, such as subcontracting specialized tasks to experienced subcontractors, they mitigate risks associated with specific project elements.

6.3 Developing Risk Mitigation Plans

Once risks are identified and assessed, construction businesses need to develop risk mitigation plans to proactively address and reduce the likelihood or impact of potential risks. Risk mitigation plans outline specific actions, procedures, and controls to minimize risks and ensure project and business continuity.

Mitigation strategies may include implementing safety protocols, conducting regular inspections, using advanced technology for risk monitoring, implementing quality control measures, and establishing effective communication channels among project stakeholders. It is essential to allocate resources and responsibilities to ensure the successful implementation of mitigation measures.

Case Study: A construction project involves working in a high-risk environment with potential safety hazards. The project team develops a comprehensive risk mitigation plan, which includes regular safety training sessions, stringent adherence to safety regulations, provision of personal protective equipment, and daily safety briefings. By implementing these measures, they reduce the likelihood of accidents, promote a safety culture, and ensure a safe working environment for all project participants.

6.4 Crisis Management and Business Continuity

Effective crisis management and business continuity planning are crucial for construction businesses to respond swiftly and efficiently to unexpected events that could disrupt project progress or business operations. Crisis management involves the implementation of protocols, procedures, and communication strategies to handle emergencies or critical incidents.

Business continuity planning focuses on developing strategies to ensure that critical business functions can continue operating during and after a crisis. This includes developing contingency plans, establishing backup systems, identifying alternative suppliers, and maintaining robust communication networks.

Case Study: A construction company experiences a significant natural disaster that severely impacts ongoing projects. Through effective crisis management, they activate emergency response protocols, evacuate workers safely, and secure the construction sites to prevent further damage. Simultaneously, their business continuity plan comes into effect, ensuring the continuation of essential business functions, such as project documentation, financial operations, and client communication, even during the crisis period.

6.5 Learning from Risk: Lessons for Business Improvement

The final aspect of risk management in the construction business involves learning from risks and using those lessons to drive continuous improvement. By analyzing past risks, their causes, and their impacts, construction businesses can identify areas for improvement, refine their processes, and strengthen their risk management strategies.

Regular review and analysis of risk management practices help identify recurring patterns, emerging risks, and opportunities for innovation. Construction businesses can leverage these insights to develop best practices, enhance decision-making frameworks, and foster a culture of continuous learning and improvement.

Case Study: A construction company encounters a series of cost overruns and delays across multiple projects. Through a comprehensive analysis of the underlying causes, they identify common issues such as inaccurate project estimation, poor subcontractor management, and insufficient project controls. By implementing corrective measures, such as refining estimation methods, improving subcontractor selection processes, and enhancing project monitoring systems, they enhance their risk management practices and significantly improve project outcomes.

By embracing robust risk management practices, professionals can navigate the complexities of the construction industry and safeguard their businesses from potential risks. Through effective identification and assessment of risks, implementation of insurance and risk transfer strategies, development of risk mitigation plans, establishment of crisis management and business continuity protocols, and continuous learning from risk, construction businesses can enhance their resilience, protect their investments, and ensure sustainable growth in a dynamic and challenging industry.

CHAPTER 7: MARKETING AND SALES

Effective marketing and sales strategies are essential for construction businesses to position themselves in the market, generate leads, build strong customer relationships, win contracts, and forge strategic partnerships. This chapter explores key aspects of marketing and sales specific to the construction industry, including branding and positioning, developing a marketing strategy, lead generation and customer relationship management, bidding and proposal preparation, and building strategic partnerships and alliances.

7.1 Branding and Positioning in Construction Business

Branding and positioning play a crucial role in establishing a construction business's identity, reputation, and differentiation in the marketplace. By developing a strong brand and effectively positioning themselves, construction businesses can attract and retain customers, foster trust, and gain a competitive edge.

Successful branding in the construction industry involves defining a clear brand identity that aligns with the company's values, mission, and unique strengths. This includes creating a compelling brand story, designing a visually appealing and professional brand image, and consistently communicating the brand message across various channels.

Positioning involves identifying the target market segments, understanding customer needs and preferences, and positioning the business as the preferred solution provider. It requires conducting market research, analyzing competitors, and identifying the unique value proposition that sets the business apart from others in the industry.

Case Study: A construction company specializing in sustainable building practices establishes a strong brand identity by highlighting its commitment to eco-friendly construction methods, energy efficiency, and environmentally conscious materials. Through targeted marketing campaigns, they position themselves as a trusted provider of sustainable construction solutions, attracting environmentally conscious clients who prioritize sustainability in their projects.

7.2 Developing a Marketing Strategy

A well-defined marketing strategy is vital for construction businesses to effectively promote their services, reach their target audience, and achieve their business objectives. Developing a comprehensive marketing strategy involves careful planning, analysis, and execution of various marketing tactics.

The marketing strategy should include a clear definition of target markets, specific marketing objectives, and a detailed plan for reaching and engaging potential customers. It should encompass both online and offline marketing channels and leverage various tactics such as digital marketing, content marketing, social media marketing, and traditional advertising methods.

Furthermore, construction businesses should consider the unique aspects of their industry, such as long sales cycles, relationship-driven sales, and the importance of referrals and testimonials. These factors should be integrated into the marketing strategy to maximize its effectiveness.

Case Study: A construction company that specializes in luxury residential projects develops a marketing strategy focused on targeting high-net-worth individuals and affluent neighborhoods. They leverage digital marketing channels, such as targeted online advertising and social media campaigns, to raise awareness of their brand and showcase their portfolio of high-end projects. Additionally, they establish relationships with local architects, interior designers, and real estate agents to gain referrals and expand their network within the luxury market segment.

7.3 Lead Generation and Customer Relationship Management

Lead generation and customer relationship management (CRM) are critical for sustaining a healthy pipeline of prospective clients and nurturing strong customer relationships. Construction businesses must implement effective lead generation strategies and adopt CRM tools to track and manage customer interactions.

Lead generation involves identifying and attracting potential customers who are likely to have construction needs or projects in the future. This can be achieved through various channels, such as targeted marketing campaigns, industry events and trade shows, referrals, and partnerships.

Once leads are generated, construction businesses should implement a robust CRM system to capture and manage customer information, track communication history, and streamline sales processes. This enables personalized and timely follow-up, efficient proposal preparation, and effective relationship building.

Case Study: A construction company implements a lead generation strategy that includes attending industry conferences and networking events, hosting educational webinars, and partnering with architectural firms to gain access to potential clients. They integrate a CRM system that tracks lead interactions, stores client preferences, and provides reminders for follow-up actions. Through systematic lead nurturing and personalized communication, they enhance customer engagement and increase their conversion rate.

7.4 Bidding and Proposal Preparation

Bidding and proposal preparation are essential elements of the construction business's sales process. Construction projects are often awarded through competitive bidding processes, requiring businesses to submit well-crafted proposals that highlight their capabilities, expertise, and value proposition.

To prepare competitive bids and proposals, construction businesses should conduct thorough project analysis, accurately estimate costs, and present their unique strengths and qualifications. This involves understanding project requirements, conducting site visits, collaborating with subcontractors and suppliers, and developing a detailed project plan.

Additionally, businesses should focus on effectively communicating their competitive advantages, such as specialized expertise, track record of successful projects, and commitment to quality and safety. Proposals should be tailored to the specific project and client, addressing their unique needs and concerns.

Case Study: A construction company receives a request for proposal (RFP) for a large-scale commercial project. They carefully analyze the RFP requirements, conduct site visits, and collaborate with relevant subcontractors and suppliers to gather accurate cost estimates. Their proposal highlights their experience in delivering similar projects on time and within budget, their expertise in the specific industry sector, and their commitment to sustainable construction practices. Through a compelling and well-structured proposal, they secure the contract over their competitors.

7.5 Building Strategic Partnerships and Alliances

Building strategic partnerships and alliances is an effective strategy for construction businesses to expand their capabilities, reach new markets, and enhance their competitive advantage. By collaborating with complementary businesses or industry stakeholders, construction companies can access additional resources, expertise, and market opportunities.

Strategic partnerships can take various forms, such as joint ventures, subcontracting agreements, or collaborative agreements with suppliers, subcontractors, architectural firms, engineering consultants, or real estate developers. These partnerships can offer synergies in terms of specialized skills, shared resources, and increased market access.

Successful partnerships rely on mutual trust, clear expectations, and effective communication. Construction businesses should carefully evaluate potential partners, establish formal agreements, and maintain ongoing collaboration and relationship management.

Case Study: A construction company forms a strategic partnership with an architectural firm that specializes in sustainable design. Through this partnership, they jointly market their integrated services, offering clients a seamless and sustainable construction experience. By combining their respective expertise and sharing resources, they expand their market reach, attract environmentally conscious clients, and deliver high-quality projects that align with sustainability standards.

Through effective implementation of marketing and sales strategies tailored to the construction industry, professionals can position their businesses for success. Through branding and positioning, developing a comprehensive marketing strategy, generating leads and managing customer relationships, preparing competitive bids and proposals, and building strategic partnerships, construction businesses can differentiate themselves, attract clients, and thrive in the dynamic construction landscape.

CHAPTER 8: HUMAN RESOURCES MANAGEMENT

Human resources management is a vital component of running a successful construction business. This chapter delves into the key aspects of managing human resources in the construction industry, including workforce planning and recruitment strategies, training and development, performance management and employee engagement, leadership and team building, and workplace diversity and inclusion.

8.1 Workforce Planning and Recruitment Strategies

Workforce planning and recruitment strategies are crucial for construction businesses to ensure they have the right talent in place to meet project demands and achieve organizational objectives. Effective workforce planning involves forecasting future labor needs, identifying skill gaps, and developing strategies to attract and retain skilled workers.

Construction businesses should conduct a comprehensive assessment of their workforce requirements based on upcoming projects, geographical locations, and specialized skill sets. This analysis helps determine whether the current workforce can fulfill the project demands or if additional recruitment is necessary.

Recruitment strategies should consider a diverse range of sourcing channels, such as job boards, industry associations, trade schools, apprenticeship programs, and networking events. Construction businesses should also establish relationships with local educational institutions and vocational training centers to tap into emerging talent pools.

Case Study: A construction company specializing in infrastructure projects anticipates a surge in demand for civil engineers due to several upcoming projects. They proactively collaborate with local universities to establish internship programs and offer mentorship opportunities. By nurturing relationships with students, they successfully attract and retain talented graduates, ensuring a skilled workforce to meet project requirements.

8.2 Training and Development in Construction

Training and development programs are essential to enhance the skills, knowledge, and competencies of construction workers. Effective training ensures that employees have the necessary expertise to perform their jobs safely, efficiently, and in compliance with industry standards.

Construction businesses should design and implement comprehensive training programs that cover various aspects, such as construction techniques, equipment operation, safety protocols, quality control, and project management. These programs should encompass both theoretical and practical components, leveraging a combination of classroom training, on-the-job training, and virtual learning platforms.

Investing in the professional development of employees not only improves their individual capabilities but also contributes to the overall success of the construction business. Continuous learning opportunities, certifications, and skill advancement programs help employees stay updated with industry trends and best practices.

Case Study: A construction company establishes a robust training and development program that includes safety training modules, equipment operation workshops, and project management courses. They leverage a combination of in-house trainers, external subject matter experts, and e-learning platforms to provide comprehensive training opportunities. As a result, their workforce demonstrates a high level of competency, adherence to safety standards, and efficient project execution.

8.3 Performance Management and Employee Engagement

Effective performance management and employee engagement practices are crucial for fostering a positive work environment, motivating employees, and maximizing productivity in the construction industry.

Performance management involves setting clear performance expectations, providing regular feedback, and conducting performance evaluations. Construction businesses should establish performance metrics and key performance indicators (KPIs) that align with organizational goals and project objectives. Regular performance reviews enable constructive discussions about strengths, areas for improvement, and career development opportunities.

Employee engagement initiatives are designed to promote a sense of belonging, job satisfaction, and commitment among construction workers. Engaged employees are more likely to go the extra mile, contribute innovative ideas, and maintain a high level of productivity. Engagement strategies may include recognition and reward programs, team-building activities, and opportunities for employee involvement in decision-making processes.

Case Study: A construction company implements a performance management system that includes quarterly performance reviews, goal-setting exercises, and skill development plans for each employee. They also establish a recognition program that acknowledges exceptional performance, safety achievements, and innovation. Through these initiatives, they foster a culture of continuous improvement, high performance, and employee satisfaction.

8.4 Leadership and Team Building in Construction

Strong leadership and effective team building are critical for successful project execution and achieving organizational goals in the construction industry. Leaders play a pivotal role in setting the vision, guiding teams, making strategic decisions, and ensuring project outcomes meet client expectations.

Construction businesses should invest in leadership development programs to nurture and empower future leaders. These programs can include leadership workshops, mentoring programs, and opportunities for cross-functional experiences. Leaders in the construction industry must possess a deep understanding of construction processes, technical expertise, and excellent communication and interpersonal skills.

Effective team building fosters collaboration, communication, and synergy among construction teams. Construction businesses should focus on creating a positive work culture that promotes teamwork, mutual respect, and shared goals. Regular team-building activities, such as workshops, off-site retreats, and problem-solving exercises, can help improve team dynamics and enhance project outcomes.

Case Study: A construction company promotes leadership development by offering management training programs and mentorship opportunities for high-potential employees. They also prioritize team building by organizing regular team-building workshops, fostering open communication channels, and encouraging collaboration among cross-functional teams. As a result, they have strong leaders who inspire their teams and drive project success.

8.5 Workplace Diversity and Inclusion

Embracing workplace diversity and inclusion is essential for construction businesses to tap into a wide range of perspectives, experiences, and talents. A diverse and inclusive workforce fosters innovation, creativity, and improved decision-making, leading to a competitive advantage in the construction industry.

Construction businesses should implement policies and practices that promote diversity and inclusion at all levels of the organization. This includes fair hiring practices, equal opportunity for career advancement, and creating an inclusive work environment that values and respects individual differences.

Promoting diversity and inclusion also involves providing training and awareness programs to educate employees about unconscious bias, cultural competence, and inclusive communication. By cultivating a diverse workforce, construction businesses can better understand and meet the needs of diverse client bases and contribute to social responsibility.

Case Study: A construction company proactively recruits and hires individuals from diverse backgrounds, including different genders, ethnicities, and cultures. They implement diversity training programs that educate employees on the benefits of diversity, inclusion, and respectful communication. As a result, they have a dynamic workforce that brings diverse perspectives to problem-solving and fosters innovation.

Construction professionals can build a skilled and engaged workforce, enhance performance and productivity, and create a positive work environment. Workforce planning and recruitment strategies, training and development programs, performance management and employee engagement, leadership and team building, and workplace diversity and inclusion are key areas that contribute to the success of construction businesses.

PART III: OPERATIONAL EXCELLENCE IN CONSTRUCTION BUSINESS

CHAPTER 9: PROJECT ESTIMATION AND COST CONTROL

Accurate project estimation and effective cost control are essential for achieving operational excellence in the construction business. This chapter explores the various methods of cost estimation in construction, cost control techniques and value engineering, budgeting and financial management, change management and variation control, as well as supplier and subcontractor management.

9.1 Methods of Cost Estimation in Construction

Cost estimation is a crucial process that involves assessing the expenses associated with a construction project. Accurate cost estimation enables construction businesses to determine the financial feasibility of a project, establish competitive bids, and allocate resources effectively. Several methods are used in the construction industry to estimate costs, including:

Quantity Takeoff: This method involves quantifying the materials, labor, and equipment required for each project component. Quantity takeoff relies on detailed project plans, specifications, and historical data to determine the quantities needed and their associated costs.

Unit Cost Estimation: Unit cost estimation involves assigning costs to standard units of work, such as per square meter, per cubic yard, or per linear foot. Construction businesses can refer to historical data, industry benchmarks, and current market rates to establish unit costs.

Parametric Estimation: Parametric estimation utilizes mathematical models and statistical analysis to estimate costs based on project parameters. These models consider factors such as project size, complexity, location, and historical data to generate cost estimates.

Bottom-Up Estimation: Bottom-up estimation involves breaking down a project into smaller components and estimating costs for each individual item. This method provides a detailed and accurate estimate but requires a thorough understanding of project requirements and extensive data analysis.

Case Study: A construction company utilizes a combination of quantity takeoff and unit cost estimation to estimate the costs for a new residential development project. By analyzing the architectural plans, determining the quantities of materials needed, and applying unit costs based on market rates, they produce a comprehensive cost estimate for the entire project.

9.2 Cost Control Techniques and Value Engineering

Cost control techniques and value engineering are essential for managing project expenses, optimizing resource utilization, and ensuring project profitability. Construction businesses employ various strategies to control costs effectively and enhance value, such as:

Cost Tracking and Reporting: Implementing robust cost tracking systems allows construction businesses to monitor project expenses in real-time. Regular financial reports enable project managers to identify cost overruns, track variances, and take corrective measures.

Value Engineering: Value engineering involves analyzing project components and processes to identify opportunities for cost reduction without compromising quality. By examining alternative materials, construction methods, and design modifications, construction businesses can achieve cost savings while maintaining or improving project outcomes.

Efficient Resource Allocation: Effective resource management is critical for cost control. Construction businesses should optimize the allocation of labor, equipment, and materials to minimize waste and enhance productivity. Proper scheduling and coordination can prevent delays, reduce idle time, and improve overall project efficiency.

Supplier and Subcontractor Evaluation: Engaging reliable suppliers and subcontractors can contribute to cost control efforts. Construction businesses should evaluate potential partners based on their track record, capabilities, and competitive pricing. Maintaining strong relationships with suppliers and subcontractors can lead to favorable pricing, timely delivery, and enhanced cost control.

Case Study: A construction company implements a value engineering approach during the design phase of a commercial building project. Through collaboration with architects, engineers, and subcontractors, they identify cost-saving opportunities, such as using energy-efficient materials, optimizing space utilization, and implementing innovative construction techniques. As a result, they achieve significant cost reductions while delivering a high-quality project to the client.

9.3 Budgeting and Financial Management

Effective budgeting and financial management are essential for the success of construction projects. Proper financial planning ensures that project expenses are adequately accounted for, cash flow is maintained, and financial risks are mitigated. Key considerations for budgeting and financial management in the construction business include:

Establishing Realistic Budgets: Construction businesses must develop accurate and comprehensive budgets that encompass all project costs, including labor, materials, equipment, subcontractors, permits, and contingencies. Budgets should account for potential risks and unforeseen circumstances to prevent cost overruns.

Cash Flow Management: Construction projects typically involve staggered payments, and effective cash flow management is crucial to maintain the financial health of the business. Construction businesses should monitor cash inflows and outflows, manage billing and payment processes, and maintain adequate reserves to cover expenses.

Financial Reporting and Analysis: Regular financial reporting provides insights into project performance, identifies areas of improvement, and ensures transparency with stakeholders. Analyzing financial data allows construction businesses to make informed decisions, identify cost-saving opportunities, and assess project profitability.

Risk Assessment and Mitigation: Financial risks are inherent in construction projects. Construction businesses should conduct thorough risk assessments to identify potential financial risks and develop mitigation strategies. This includes analyzing project contingencies, evaluating insurance coverage, and implementing financial controls to minimize risk exposure.

Case Study: A construction company utilizes advanced budgeting software to develop accurate project budgets and monitor financial performance in real-time. The software integrates project expenses, tracks cash flow, and generates detailed financial reports. This enables the company to proactively manage finances, identify cost variances, and make data-driven decisions to ensure project profitability.

9.4 Change Management and Variation Control

Construction projects often encounter changes and variations that can impact costs and project timelines. Effective change management and variation control processes are essential to minimize the impact of changes on project budgets and schedules. Key considerations for change management and variation control include:

Change Order Management: Construction businesses should establish robust change order management procedures to assess the impact of changes on project costs, schedules, and resources. This involves evaluating change requests, negotiating pricing adjustments, documenting changes, and obtaining client approvals.

Variation Tracking and Reporting: Maintaining a comprehensive record of variations throughout the project lifecycle is crucial. Construction businesses should track and report variations, including the associated costs, to provide transparency and facilitate accurate financial analysis.

Documentation and Communication: Clear documentation and effective communication are vital for managing changes and variations. Construction businesses should maintain a centralized repository of project documents, including change orders, variation requests, and related correspondence. Timely communication with stakeholders helps manage expectations and minimize disputes.

Risk Assessment and Mitigation: Changes and variations introduce risks to project budgets and schedules. Construction businesses should conduct risk assessments to identify potential risks associated with changes and develop mitigation strategies. This may include contingency planning, renegotiating contracts, or revising project plans.

Case Study: A construction company establishes a dedicated change management team responsible for evaluating change requests, estimating their impact on costs and schedules, and obtaining necessary approvals. They implement a robust documentation system to track and record all changes throughout the project. This proactive approach helps them effectively manage variations, mitigate risks, and maintain project profitability.

9.5 Supplier and Subcontractor Management

Effective supplier and subcontractor management are critical for controlling project costs, ensuring timely delivery of materials and services, and maintaining quality standards. Construction businesses should establish strong relationships with suppliers and subcontractors and implement sound management practices, including:

Supplier and Subcontractor Selection: Construction businesses should carefully evaluate potential suppliers and subcontractors based on their capabilities, reliability, track record, and pricing. Establishing long-term partnerships with trusted suppliers and subcontractors can result in preferential pricing, improved quality, and timely delivery.

Contract Management: Developing clear and comprehensive contracts with suppliers and subcontractors is essential. Contracts should include specific deliverables, pricing terms, quality standards, and project timelines. Regularly reviewing and monitoring contracts helps ensure compliance and minimizes contractual disputes.

Performance Evaluation: Regular performance evaluation of suppliers and subcontractors allows construction businesses to assess their reliability, adherence to project specifications, and overall performance. This evaluation can include factors such as on-time delivery, quality of work, and responsiveness to issues and concerns.

Communication and Collaboration: Effective communication and collaboration with suppliers and subcontractors are crucial for smooth project execution. Construction businesses should maintain open lines of communication, provide clear project requirements, and foster collaborative relationships. This facilitates proactive problem-solving and reduces the risk of delays or quality issues.

Case Study: A construction company establishes a structured supplier and subcontractor management process. They conduct thorough prequalification assessments, assess supplier and subcontractor capabilities, and negotiate contracts based on performance indicators and key deliverables. Regular performance evaluations and constructive feedback sessions foster a collaborative environment, leading to improved project outcomes and enhanced cost control.

By addressing the critical areas of project estimation, cost control, budgeting, change management, and supplier/subcontractor management, construction businesses can strive for operational excellence. These essential functions contribute to the overall success, profitability, and sustainability of construction projects.

As construction professionals, it is imperative that we understand the significance of these operational practices and their direct impact on the success of construction projects.

CHAPTER 10: QUALITY MANAGEMENT

In the construction industry, ensuring high-quality standards is of paramount importance to achieve successful project outcomes and maintain client satisfaction. Quality management encompasses a range of processes and practices aimed at planning, controlling, and improving the quality of construction projects. This chapter delves into the key aspects of quality management in the context of the construction business.

10.1 Quality Planning and Standards

Quality planning is the foundation for achieving desired project outcomes and delivering a product that meets or exceeds client expectations. It involves establishing quality objectives, defining quality standards, and developing a comprehensive quality management plan. Key considerations for quality planning in construction include:

Defining Quality Objectives: Construction businesses should clearly define their quality objectives, which may encompass aspects such as adherence to technical specifications, compliance with regulations and codes, and meeting project milestones. Quality objectives should align with the overall project goals and client requirements.

Setting Quality Standards: Establishing quality standards is essential to ensure consistency and uniformity in project deliverables. These standards may cover areas such as materials, workmanship, safety practices, and environmental sustainability. Standards should be based on industry best practices, relevant regulations, and client expectations.

Quality Control Measures: Quality control involves the systematic inspection, testing, and verification of construction processes and outputs. Construction businesses should develop robust quality control procedures to monitor and validate adherence to established standards. This may include regular inspections, sampling and testing of materials, and performance evaluations.

Non-Conformance Management: Construction projects may encounter non-conformances or deviations from quality standards. Effective non-conformance management involves identifying, documenting, and addressing instances of non-compliance. It includes implementing corrective and preventive actions to rectify issues and prevent recurrence.

Case Study: A construction company implements a quality planning process for a residential building project. They define quality objectives such as achieving zero defects, meeting project deadlines, and ensuring client satisfaction. The company establishes quality standards based on local building codes, industry best practices, and energy efficiency guidelines. Quality control measures, including regular inspections, third-party testing of materials, and internal audits, are implemented to monitor compliance and maintain quality throughout the project lifecycle.

10.2 Quality Control and Assurance

Quality control and assurance are integral components of an effective quality management system in construction. While quality control focuses on identifying and addressing non-conformances during project execution, quality assurance encompasses proactive measures to prevent quality issues. Key considerations for quality control and assurance in construction include:

Inspection and Testing: Regular inspections and testing of materials, components, and construction processes are essential for quality control and assurance. Construction businesses should establish inspection protocols, sampling plans, and testing procedures to verify compliance with quality standards. This may involve visual inspections, destructive and non-destructive testing, and performance evaluations.

Quality Audits: Conducting regular internal and external quality audits helps assess the effectiveness of quality management systems and processes. Audits identify areas of improvement, non-compliance, and potential risks. Construction businesses should develop comprehensive audit plans, assign qualified auditors, and implement corrective actions based on audit findings.

Document Control: Proper document control ensures that accurate and up-to-date documentation, including specifications, drawings, procedures, and records, is maintained throughout the project. Document control systems facilitate version control, accessibility, and traceability, which are crucial for quality control and assurance.

Supplier and Subcontractor Quality Management: Construction businesses should establish robust quality management processes for suppliers and subcontractors. This involves evaluating their quality systems, conducting supplier audits, and monitoring their performance. Close collaboration with suppliers and subcontractors helps ensure that quality requirements are met throughout the supply chain.

Case Study: A construction company implements a

comprehensive quality control and assurance program for a large-scale infrastructure project. They conduct regular inspections of critical structural components, perform material testing in accredited laboratories, and maintain a centralized document control system. Internal quality audits are carried out at various project stages, identifying areas for improvement and ensuring compliance with quality standards. The company also establishes a supplier evaluation process, including audits and performance monitoring, to maintain quality consistency in the procurement process.

10.3 Implementing a Quality Management System

To achieve consistent and effective quality management, construction businesses should establish a quality management system (QMS). A QMS provides a structured framework for planning, implementing, and monitoring quality-related activities. Key elements of implementing a QMS include:

Quality Policy: Developing a quality policy statement that defines the organization's commitment to quality and serves as a guiding principle for all quality-related activities.

Process Mapping: Identifying and documenting key processes involved in the project lifecycle, including design, procurement, construction, and inspection. Process mapping helps understand the interdependencies between activities and ensures consistency in quality management across the organization.

Standard Operating Procedures (SOPs): Developing SOPs that outline step-by-step instructions for performing specific tasks and processes. SOPs provide clarity, ensure uniformity, and serve as a reference for employees involved in quality-related activities.

Training and Competency Development: Providing training programs to enhance employees' knowledge and skills in quality management. This includes training on quality control techniques, inspection methods, and compliance with quality standards. Ensuring employees' competency in quality management contributes to consistent and reliable project outcomes.

Documented Procedures and Forms: Establishing a centralized repository for quality-related documents, including procedures, forms, checklists, and templates. This facilitates easy access, version control, and traceability of quality documentation.

Case Study: A construction company adopts a QMS based on international quality standards for their commercial building projects. They develop a quality policy emphasizing customer satisfaction, continuous improvement, and compliance with relevant regulations. Process mapping exercises identify critical quality control points and determine responsibilities for each process. SOPs are created for activities such as material receiving, concrete testing, and final inspections. The company conducts regular training sessions to enhance employees' understanding of the QMS and ensure consistent implementation.

10.4 Continuous Improvement and Benchmarking

Continuous improvement is an essential aspect of quality management in construction. By continually analyzing performance, identifying areas for improvement, and implementing corrective actions, construction businesses can enhance their processes and deliver even higher levels of quality. Benchmarking against industry standards and best practices further aids in identifying opportunities for improvement. Key considerations for continuous improvement and benchmarking in construction include:

Data Collection and Analysis: Collecting relevant data on quality performance indicators, such as defect rates, rework costs, and customer feedback. Analyzing this data helps identify trends, root causes of quality issues, and areas for improvement.

Corrective and Preventive Actions: Implementing corrective actions to address non-conformances and prevent recurrence of quality issues. Concurrently, preventive actions focus on identifying potential risks and taking proactive measures to mitigate them before they impact project quality.

Lessons Learned: Capturing lessons learned from previous projects and incorporating them into current practices. This knowledge sharing ensures that valuable insights and best practices are applied to future projects, contributing to continuous improvement.

Benchmarking: Comparing performance metrics, processes, and practices against industry benchmarks and best-in-class organizations. Benchmarking helps identify areas of excellence and areas where improvements can be made. Construction businesses can adapt successful practices and methodologies to enhance their own quality management approaches.

Case Study: A construction company adopts a continuous improvement approach for their infrastructure projects. They establish a robust data collection and analysis system, capturing information on defects, rework costs, and client feedback. Root cause analysis is performed to identify the underlying reasons for quality issues, leading to the implementation of corrective actions. Regular project review meetings are conducted to share lessons learned and identify improvement opportunities. The company also engages in benchmarking exercises to compare their quality performance against industry standards, implementing changes to align with best practices.

10.5 Customer Satisfaction

and Feedback

The satisfaction of clients and stakeholders is a crucial measure of quality in the construction industry. By actively seeking feedback and incorporating it into their quality management processes, construction businesses can enhance customer satisfaction and strengthen relationships. Key considerations for managing customer satisfaction and feedback in construction include:

Customer Engagement: Engaging with clients and stakeholders throughout the project lifecycle to understand their expectations, address concerns, and provide regular updates. Open communication channels foster transparency and trust, leading to improved satisfaction.

Post-Project Evaluation: Conducting post-project evaluations to gather feedback on project quality, timeliness, communication, and overall experience. Feedback can be collected through surveys, interviews, or focus groups. Analyzing this feedback helps identify areas of improvement and further strengthens the quality management approach.

Continuous Communication: Establishing mechanisms for ongoing communication with clients and stakeholders to address any quality-related issues promptly. This may include designated points of contact, regular progress updates, and forums for feedback and resolution.

Lessons Learned and Best Practices Sharing: Sharing success stories, lessons learned, and best practices with clients and stakeholders. Demonstrating a commitment to quality and continuous improvement builds confidence and trust in the construction business's capabilities.

Case Study: A construction company places significant emphasis on customer satisfaction and feedback for their residential projects. They conduct regular meetings with clients to understand their preferences and expectations, ensuring that project design and execution align with their vision. Post-project evaluations are conducted, and feedback is collected through structured surveys and face-to-face interviews. The company establishes a dedicated customer support team to promptly address any concerns or issues raised by clients. Lessons learned from each project are shared with clients and integrated into future projects to enhance overall quality and customer satisfaction.

By focusing on quality planning, implementing effective control and assurance measures, establishing a robust quality management system, continuously improving processes, and prioritizing customer satisfaction, construction businesses can drive operational excellence and differentiate themselves in the competitive marketplace.

CHAPTER 11: TECHNOLOGY AND INNOVATION

11.1 Digital Transformation in Construction

In the ever-evolving construction industry, digital transformation has emerged as a key driver of change. It encompasses the integration of digital technologies and processes to revolutionize traditional construction practices. Digital transformation enables construction businesses to streamline operations, improve productivity, enhance collaboration, and achieve better project outcomes.

One aspect of digital transformation is the adoption of cloud computing. Cloud-based platforms provide construction companies with secure and accessible storage for project data, allowing seamless collaboration among team members. Cloud technology also facilitates real-time information sharing, enabling stakeholders to work together efficiently, regardless of their physical location. With cloud computing, project documentation, plans, and specifications can be easily accessed and updated, ensuring that all team members are working with the latest information.

Mobile applications have become indispensable tools in the construction industry. These applications enable field personnel to access project-related information, such as drawings, schedules, and specifications, directly from their mobile devices. Field teams can capture and report data, communicate progress, and address issues in real time, enhancing productivity and reducing delays. Mobile applications also support digital inspections, safety reporting, and quality control, allowing for faster and more accurate data collection.

Data analytics and artificial intelligence (AI) have transformed how construction companies analyze and leverage project information. By harnessing data from various sources, such as sensors, drones, and construction management systems, construction businesses can gain valuable insights into project performance, identify trends, and make data-driven decisions. AI-powered algorithms can analyze vast amounts of data to identify patterns, optimize processes, and improve project efficiency. For example, AI can assist in predicting equipment maintenance needs, optimizing resource allocation, or even enhancing safety protocols.

11.2 Building Information Modeling (BIM)

Building Information Modeling (BIM) has revolutionized the way construction projects are designed, planned, and executed. BIM is a collaborative process that involves creating and managing digital representations of the physical and functional aspects of a building. The 3D models generated through BIM integrate architectural, structural, and MEP systems, providing a comprehensive and coordinated view of the project.

BIM enhances design coordination by allowing different disciplines to work concurrently and identify clashes or conflicts early in the design phase. By detecting clashes in the virtual environment, BIM helps reduce rework and costly on-site modifications. The coordination and clash detection capabilities of BIM enable project teams to resolve conflicts before construction begins, saving time and resources.

The visual nature of BIM models enhances communication and understanding among project stakeholders. Clients, architects, engineers, and contractors can visualize the project in 3D, gaining a clear understanding of the design intent and facilitating effective communication throughout the project lifecycle. BIM models also support visualization tools such as virtual reality (VR) and augmented reality (AR), allowing stakeholders to experience a virtual walkthrough of the project before construction begins.

Furthermore, BIM contributes to the overall efficiency and accuracy of construction projects. BIM models provide a centralized repository of project information, including specifications, quantities, and construction schedules. This centralized data enables accurate quantity takeoffs, precise cost estimation, and improved project scheduling. BIM also supports the generation of construction documentation, including detailed drawings, schedules, and material lists, leading to better construction coordination and execution.

11.3 Construction Software and Automation

The advancements in construction software and automation have significantly impacted the efficiency and productivity of construction projects. Construction software encompasses a wide range of applications designed to streamline various processes, including project management, scheduling, cost estimation, and document control.

Project management software provides construction businesses with comprehensive tools to plan, execute, and monitor projects effectively. These software solutions facilitate project scheduling, task allocation, resource management, and collaboration among team members. They offer features such as Gantt charts, critical path analysis, and progress tracking, allowing project managers to effectively manage project timelines and ensure timely completion.

Scheduling software enables construction companies to develop accurate and optimized project schedules. These tools automate the scheduling process, taking into account dependencies, resource availability, and project constraints. Construction scheduling software allows for efficient resource allocation, minimizes delays, and enhances overall project coordination.

Cost estimation software plays a critical role in ensuring accurate and competitive project pricing. By leveraging historical data, industry benchmarks, and cost databases, construction companies can generate reliable cost estimates for different project components. Cost estimation software facilitates accurate quantity takeoffs, labor cost calculations, and material cost assessments, enabling construction businesses to develop realistic and profitable bids.

Document control software provides a centralized platform for storing, organizing, and managing project documentation. These solutions enable version control, access controls, and document tracking, ensuring that all stakeholders have access to the most up-to-date project information. Document control software improves collaboration, reduces errors arising from outdated information, and enhances overall project documentation management.

Automation technologies, such as robotics and drones, are also gaining prominence in the construction industry. Robotic systems can automate repetitive and labor-intensive tasks, such as bricklaying, concrete pouring, and material handling. Drones equipped with cameras and sensors allow for aerial inspections, surveying, and progress monitoring. These automation technologies enhance productivity, improve accuracy, and mitigate safety risks on construction sites.

By incorporating construction software and automation, construction businesses can streamline processes, reduce manual errors, and optimize resource utilization. These technologies enhance collaboration, improve project communication, and ultimately contribute to efficient project delivery.

11.4 Smart Construction and Internet of Things (IoT)

Smart construction refers to the integration of IoT devices and systems into the construction process to create intelligent and interconnected construction environments. IoT devices, equipped with sensors and communication capabilities, enable the collection and exchange of data in real time. By leveraging IoT technology, construction businesses can enhance safety, monitor site conditions, optimize resource usage, and improve project outcomes.

One significant application of IoT in construction is the real-time monitoring of construction sites. IoT sensors can be installed to monitor various parameters, such as temperature, humidity, vibration, and air quality. Real-time monitoring enables early detection of potential hazards, such as gas leaks, excessive dust levels, or structural instability. With this data, construction companies can implement proactive measures to mitigate risks and ensure the safety of workers on site.

IoT devices can also monitor and optimize the usage of construction resources, such as energy and water. Smart meters and sensors enable real-time monitoring of energy consumption, allowing construction businesses to identify areas of inefficiency and implement energy-saving measures. Water sensors can detect leaks or excessive water usage, enabling prompt action and conservation efforts. By optimizing resource usage, construction companies can reduce costs and minimize environmental impact.

The integration of IoT devices with construction equipment and machinery enhances equipment management and maintenance. IoT-enabled equipment can collect data on usage patterns, performance, and maintenance requirements. By monitoring equipment health in real time, construction businesses can schedule maintenance proactively, reducing unplanned downtime and extending the lifespan of equipment.

Furthermore, IoT devices facilitate improved project tracking and logistics management. GPS tracking and RFID technologies allow for real-time tracking of construction materials, equipment, and personnel. Construction companies can optimize material deliveries, monitor equipment utilization, and ensure that workers are deployed efficiently on-site. This level of tracking and visibility enhances project coordination, reduces delays, and improves overall project efficiency.

11.5 Sustainable Construction Practices

Sustainable construction practices have gained significant traction in recent years, driven by the need to mitigate environmental impact and address climate change concerns. Construction businesses are increasingly adopting sustainable practices that focus on reducing energy consumption, minimizing waste generation, and using environmentally friendly materials.

Energy-efficient design and construction strategies play a crucial role in sustainable construction. Building envelopes are designed to maximize insulation, reduce thermal bridging, and optimize natural lighting and ventilation. Energy-efficient HVAC systems, lighting fixtures, and renewable energy technologies, such as solar panels or wind turbines, are incorporated into building designs to minimize energy consumption and reliance on fossil fuels.

Waste management and recycling are integral components of sustainable construction. Construction companies implement waste reduction strategies by prioritizing materials with high recycling potential, minimizing material waste during construction, and segregating waste streams on-site. Recycled materials, such as reclaimed timber, recycled concrete, or recycled steel, are utilized whenever possible to reduce the demand for virgin resources.

Water conservation is another key focus area in sustainable construction. Water-efficient fixtures, such as low-flow toilets and faucets, are installed to minimize water usage in buildings. Rainwater harvesting systems capture and reuse rainwater for irrigation or non-potable purposes, reducing reliance on municipal water sources.

The selection of environmentally friendly materials is an essential aspect of sustainable construction. Construction businesses are increasingly using sustainable and renewable materials, such as bamboo, recycled steel, and environmentally friendly insulation materials. These materials have lower embodied energy and carbon footprint, promoting greener construction practices.

CHAPTER 12: SUSTAINABILITY AND GREEN BUILDING

12.1 Understanding Sustainable Construction

In the realm of construction, sustainable practices have become increasingly essential. Sustainable construction involves adopting strategies that minimize the negative impact on the environment, conserve resources, and prioritize the health and well-being of occupants. Understanding the principles and goals of sustainable construction is crucial for construction professionals to make informed decisions and implement sustainable strategies effectively.

Sustainable construction encompasses multiple facets, including energy efficiency, water conservation, waste reduction, and the use of environmentally friendly materials. It involves designing buildings that optimize natural resources, minimize energy consumption, and reduce greenhouse gas emissions. By integrating sustainable design principles, such as passive solar design, efficient insulation, and energy-efficient lighting systems, construction professionals can create buildings that significantly reduce their environmental footprint.

12.2 Green Building Certifications

and Standards

Green building certifications and standards provide a comprehensive framework for assessing and recognizing sustainable construction practices. Certifications like LEED (Leadership in Energy and Environmental Design) and BREEAM (Building Research Establishment Environmental Assessment Method) establish benchmarks and criteria for sustainable building design, construction, and operation. These certifications consider factors such as energy efficiency, water conservation, indoor air quality, and materials selection.

By pursuing green building certifications, construction businesses demonstrate their commitment to sustainability and gain recognition for their efforts. Certifications not only enhance the marketability of buildings but also contribute to energy savings, reduced environmental impact, and improved occupant health and comfort. They provide a standardized approach to sustainable construction and serve as a guide for implementing best practices throughout the project lifecycle.

12.3 Energy Efficiency and Renewable Energy

Energy efficiency plays a pivotal role in sustainable construction. Construction professionals should employ energy-efficient design strategies and technologies to reduce energy consumption in buildings. This includes using effective insulation, high-performance windows, and efficient heating, ventilation, and air conditioning (HVAC) systems. By incorporating smart building technologies, such as occupancy sensors and energy management systems, energy usage can be optimized further.

In addition to energy efficiency, renewable energy technologies offer opportunities to generate clean energy on-site. Installing solar panels, wind turbines, or geothermal systems can offset a building's energy demand and reduce reliance on fossil fuels.

Construction professionals should consider the feasibility of incorporating renewable energy systems during the design and construction phases, taking into account factors such as available resources, site conditions, and financial considerations.

12.4 Waste Management and Recycling

Effective waste management and recycling practices are crucial for sustainable construction. Construction sites generate significant amounts of waste, including construction debris, packaging materials, and excess materials. Construction businesses should implement waste management plans that prioritize waste reduction, reuse, and recycling to minimize environmental impact and promote resource conservation.

Waste reduction strategies include optimizing material usage, implementing construction waste management plans, and segregating waste streams on-site. Recycling initiatives should target materials such as concrete, metal, wood, and plastics. Construction businesses can partner with recycling facilities and waste management companies to ensure proper disposal and recycling of construction waste, diverting it from landfills and contributing to a circular economy.

12.5 Incorporating Sustainability into Business Strategy

To truly embrace sustainability, construction businesses need to integrate it into their overall business strategy. This involves aligning sustainability goals with core business objectives, adopting sustainable procurement practices, and fostering a culture of sustainability within the organization.

Sustainability should be considered from project inception to completion, involving all stakeholders, including architects, engineers, subcontractors, and suppliers. Construction businesses can establish sustainability committees or task forces to oversee sustainability initiatives and ensure continuous improvement. By incorporating sustainable practices into business operations, such as reducing energy and water consumption, implementing green purchasing policies, and promoting sustainable transportation, construction companies can demonstrate environmental leadership and improve their overall performance.

Construction companies can engage with clients and educate them about the benefits of sustainable construction. By promoting the value of sustainable buildings, including lower operating costs, enhanced occupant comfort, and improved environmental stewardship, construction businesses can drive demand for sustainable practices and influence industry norms.

PART IV: NAVIGATING CHALLENGES AND ENSURING SUCCESS

CHAPTER 13: PROJECT DELIVERY METHODS

13.1 Traditional and Design-Bid-Build Method

The traditional and widely used project delivery method in the construction industry is the Design-Bid-Build (DBB) method. Under this approach, the project is divided into sequential phases, beginning with the design phase, followed by the bidding or tendering phase, and concluding with the construction phase. The owner contracts separately with an architect or designer to develop the project design and with a general contractor through a competitive bidding process.

The DBB method offers distinct advantages, such as clear separation of design and construction responsibilities and the opportunity for competitive bidding. However, it also has limitations, including potential delays due to sequential phases, limited collaboration between the designer and contractor during the design phase, and potential conflicts between the owner, designer, and contractor.

13.2 Construction Management at Risk (CMAR)

Construction Management at Risk (CMAR) is a project delivery method that involves the early involvement of a construction manager (CM) during the design phase. The CM is responsible for providing input on constructability, cost estimating, and scheduling, working collaboratively with the owner and design team. Once the design is complete, the CM assumes the role of the general contractor and manages the construction phase.

CMAR offers advantages such as early contractor involvement, improved constructability, and cost and schedule control. It promotes collaboration between the owner, designer, and construction manager, allowing for greater project efficiency and reduced risk. CMAR is particularly beneficial for complex projects or projects with tight schedules where early input from construction professionals is crucial.

13.3 Design-Build and Integrated Project Delivery (IPD)

Design-Build (DB) and Integrated Project Delivery (IPD) are collaborative project delivery methods that involve close cooperation between the owner, designer, and contractor throughout the project lifecycle. In the DB method, the owner contracts with a single entity that assumes responsibility for both design and construction. IPD takes collaboration a step further, with all stakeholders, including the owner, designer, contractor, and key subcontractors, entering into a multi-party agreement.

DB and IPD offer advantages such as enhanced communication and coordination, streamlined decision-making, and shared risk and reward among project participants. By involving all parties from the project's early stages, these methods foster a collaborative environment that promotes innovation, problem-solving, and value optimization.

13.4 Public-Private Partnerships (PPP)

Public-Private Partnerships (PPP) involve collaboration between a public entity, such as a government agency, and a private sector entity to jointly undertake a construction project. In this delivery method, the public entity typically provides the project site and certain project requirements, while the private entity assumes responsibility for design, construction, financing, operation, and maintenance.

PPP offers benefits such as increased access to private sector expertise and resources, efficient project delivery, and risk sharing between the public and private sectors. It enables the public sector to leverage private sector efficiencies and innovation while transferring certain project risks to the private partner.

13.5 Choosing the Right Project Delivery Method

Selecting the appropriate project delivery method is a critical decision for construction projects. Factors that influence the choice of delivery method include project complexity, schedule constraints, budget considerations, and the owner's risk tolerance. Each project delivery method has its strengths and limitations, and the selection should align with the project's specific needs and objectives.

Construction professionals must carefully evaluate the advantages, disadvantages, and unique characteristics of each delivery method to determine the most suitable approach. They should consider factors such as project size, complexity, budget, timeline, stakeholder requirements, and the desired level of collaboration and risk allocation. Case studies and examples can provide valuable insights into successful projects that utilized different delivery methods.

CHAPTER 14: ETHICS AND PROFESSIONALISM

14.1 Ethical Principles in Construction Business

Ethical principles form the foundation of a reputable and successful construction business. Upholding ethical standards ensures that construction professionals act with integrity, honesty, and fairness in all their business dealings. In the construction industry, ethical behavior extends to various aspects, including client relationships, project execution, employee management, and engagement with stakeholders.

Key ethical principles in construction business include transparency, accountability, respect for legal and regulatory requirements, and commitment to delivering quality work. Construction professionals should prioritize ethical conduct by adhering to professional standards, promoting a culture of integrity within their organizations, and establishing robust ethical guidelines and policies.

14.2 Professional Codes of Conduct

Professional codes of conduct serve as guidelines for construction professionals to uphold ethical behavior and professionalism in their work. These codes outline the expected standards of conduct, responsibilities, and obligations of professionals in the construction industry. They provide guidance on ethical decision-making, conflict resolution, and maintaining professional relationships.

Professional organizations and industry associations often develop and enforce codes of conduct specific to the construction sector. These codes emphasize principles such as honesty, fairness, confidentiality, avoiding conflicts of interest, and promoting safety and sustainability. Adhering to professional codes of conduct helps establish trust, credibility, and a positive reputation within the construction industry.

14.3 Ethical Decision-Making in Construction

Construction professionals frequently encounter complex situations that require ethical decision-making. Ethical decision-making involves considering the potential impact of actions on stakeholders, evaluating available options, and selecting the course of action that aligns with ethical principles and values. It requires a thoughtful analysis of the ethical implications, weighing potential consequences, and maintaining transparency and fairness.

To enhance ethical decision-making, construction professionals should develop ethical reasoning skills, engage in ethical discussions and training, and create an environment that encourages open communication. Case studies and examples can be valuable tools for exploring real-world ethical dilemmas in construction and understanding how ethical principles can guide decision-making.

14.4 Integrity in Business Relationships

Integrity is essential for building and maintaining strong business relationships in the construction industry. It involves being honest, trustworthy, and reliable in all interactions with clients, subcontractors, suppliers, and other stakeholders. Construction professionals must honor their commitments, deliver projects as promised, and communicate openly and transparently.

Integrity extends beyond contractual obligations and encompasses fair competition, avoiding conflicts of interest, and respecting intellectual property rights. Building a reputation for integrity fosters trust among stakeholders and enhances the likelihood of repeat business, referrals, and long-term partnerships.

14.5 Corporate Social Responsibility (CSR) in Construction

Corporate Social Responsibility (CSR) refers to the commitment of construction businesses to contribute positively to society and the environment. CSR in construction involves considering the social, environmental, and economic impacts of construction activities and taking proactive steps to minimize negative effects and create sustainable value.

Construction companies can demonstrate CSR through initiatives such as environmentally friendly construction practices, community engagement, employee well-being programs, and supporting local businesses and organizations. By integrating CSR into their operations, construction professionals can enhance their reputation, attract socially conscious clients, and contribute to the overall well-being of communities.

CHAPTER 15: SUCCESSION PLANNING AND BUSINESS CONTINUITY

15.1 Importance of Succession Planning

Succession planning is a critical aspect of long-term success in the construction industry. It involves identifying and developing future leaders within an organization to ensure a smooth transition of leadership and maintain business continuity. Succession planning is vital because it addresses the inevitable changes in leadership that occur due to retirement, promotions, or unexpected departures.

In the construction industry, where projects can span several years and require specialized knowledge and experience, effective succession planning is essential to preserve institutional knowledge, maintain client relationships, and sustain the company's reputation. By proactively identifying and grooming future leaders, construction businesses can minimize disruptions, mitigate risks, and position themselves for continued success.

15.2 Identifying and Developing Future Leaders

Identifying and developing future leaders is a key component of succession planning. Construction businesses must identify individuals with the potential to assume leadership roles and invest in their development. This process involves assessing current employees' skills, competencies, and leadership potential and providing them with opportunities for growth and advancement.

To identify future leaders, construction businesses can implement talent assessment programs, mentorship initiatives, and leadership development training. By identifying individuals with a combination of technical expertise, interpersonal skills, and a commitment to the company's values, organizations can nurture a pipeline of capable leaders who can guide the business forward.

15.3 Exit Strategies and Business Transition

Exit strategies and business transition plans are crucial for a seamless transfer of leadership and ownership within a construction company. Exit strategies outline the steps and considerations involved in the departure of key leaders, such as founders or senior executives. These strategies may include succession plans, buy-sell agreements, or the sale of the business to external parties.

Business transition plans ensure a smooth handover of responsibilities, knowledge transfer, and the preservation of client relationships during leadership transitions. They may involve phased retirements, mentoring arrangements, or the involvement of external consultants to facilitate the transition process. By carefully planning and executing business transitions, construction businesses can maintain stability, continuity, and client confidence.

15.4 Ensuring Business Continuity and Sustainability

Succession planning and business continuity are closely intertwined. Effective succession planning ensures the continuity and sustainability of a construction business in the face of leadership changes. Business continuity strategies involve establishing processes, systems, and safeguards to mitigate potential disruptions and maintain operations during unexpected events or leadership transitions.

Construction businesses can develop business continuity plans that outline procedures for transferring leadership, managing projects, preserving client relationships, and addressing potential risks. These plans should encompass elements such as emergency preparedness, knowledge management, and the identification of key roles and responsibilities. By proactively addressing potential disruptions, construction businesses can safeguard their reputation and minimize negative impacts on projects and stakeholders.

15.5 Learning from Successful Construction Businesses

Learning from successful construction businesses can provide valuable insights and best practices for effective succession planning and business continuity. Case studies and examples of construction companies that have navigated successful leadership transitions can offer inspiration and practical strategies.

Examining successful construction businesses can reveal the

importance of early succession planning, fostering a culture of learning and development, and embracing innovation. These examples highlight the significance of mentorship, knowledge sharing, and leveraging technology to ensure a smooth transition and sustained success.

CONCLUSION

This book offered an exceptional and comprehensive resource that encompasses every facet of the construction industry, empowering readers to navigate the complex landscape, overcome challenges, and achieve remarkable success. Throughout the book, we have explored a vast array of topics, delving deep into the intricacies of legal and regulatory frameworks, market analysis, financial management, project execution, risk mitigation, marketing and sales, human resources management, sustainability, ethics, and professionalism. We have examined the role of technology and innovation in driving transformation within the construction industry.

The book has provided invaluable insights into the legal and regulatory aspects of the construction business. By understanding the laws, regulations, and building codes that govern the industry, readers are equipped with the knowledge necessary to ensure compliance, mitigate risks, and protect their projects and stakeholders. The exploration of licensing and permitting requirements has highlighted the importance of adhering to industry standards and obtaining the necessary authorizations to operate legally and ethically.

Market analysis and business planning have been extensively covered, offering readers a strategic approach to identifying target markets, conducting market research, and developing robust business plans. By understanding market dynamics, consumer preferences, and emerging trends, construction professionals can position themselves competitively and capitalize on lucrative opportunities. Moreover, the emphasis on defining a unique selling proposition (USP) has emphasized the significance of differentiation in a crowded marketplace.

Financial management, an essential aspect of any successful business, has been meticulously addressed. From financial planning and budgeting to cash flow management and profitability analysis, readers have gained a comprehensive understanding of financial principles and practices specific to the construction industry. The exploration of cost estimation, pricing strategies, and financial statements has equipped readers with the tools to make informed financial decisions, optimize resources, and maximize profitability.

Project management, a critical discipline in construction, has been given extensive attention. By addressing project initiation and planning, organization and team building, scheduling and time management, resource allocation and procurement, and project monitoring, control, and closure, readers are equipped with the skills and knowledge necessary to execute projects efficiently, meet deadlines, and deliver exceptional results. The exploration of project management methodologies, tools, and techniques ensures that readers are well-prepared to handle the complexities and challenges inherent in construction projects.

The book has underscored the significance of risk management in the construction industry. By identifying and assessing business risks, developing risk mitigation plans, and implementing crisis management strategies, construction professionals can proactively address potential disruptions and safeguard their projects, reputation, and workforce. The emphasis on learning from risk and using those experiences for business improvement fosters a culture of continuous learning and adaptation.

Marketing and sales have been explored in depth, recognizing their pivotal role in driving business growth and success. From branding and positioning to developing comprehensive marketing strategies, lead generation, customer relationship management, and bidding and proposal preparation, readers are equipped with the tools and techniques to effectively market their construction services, build strong client relationships, and secure profitable contracts. Additionally, the exploration of building strategic partnerships and alliances highlights the importance of collaboration and synergies within the construction ecosystem.

Human resources management, a crucial component of construction business, has been thoroughly addressed. Workforce planning and recruitment strategies, training and development programs, performance management, leadership, team building, and workplace diversity and inclusion have been examined in detail. By fostering a positive and inclusive work environment, construction professionals can attract and retain top talent, nurture employee engagement, and drive organizational success.

The book has also highlighted the importance of quality management in the construction industry. By setting quality standards, implementing robust quality control and assurance measures, and fostering a culture of continuous improvement, construction professionals can deliver projects of exceptional quality, meet client expectations, and enhance their reputation. The emphasis on sustainable construction practices, green building certifications, energy efficiency, and waste management further highlights the industry's commitment to environmental stewardship and sustainable development.

Lastly, the book has addressed the significance of technology and innovation in shaping the future of the construction industry. From digital transformation and Building Information Modeling (BIM) to construction software, automation, smart construction, and the Internet of Things (IoT), readers have gained insights into the transformative power of technology in streamlining processes, improving efficiency, and enhancing project outcomes. Furthermore, the exploration of sustainable construction practices emphasizes the industry's commitment to environmental sustainability and responsible business practices.

ABOUT THE AUTHOR

Steven Smith

Steven Smith is a renowned expert in the field of Construction Management, with a wealth of knowledge and experience spanning both academia and industry. With a doctorate in Construction Management from a prestigious university, Smith has dedicated his career to advancing the field and contributing to its body of knowledge.

Smith's academic journey began with a passion for understanding the intricacies of construction processes and finding innovative solutions to the challenges faced in the industry. His doctoral research focused on optimizing project management practices and enhancing productivity in construction projects. Through rigorous study and extensive research, Smith has developed a deep understanding of the various aspects of construction management and their impact on project success.

In addition to his academic achievements, Smith has also gained significant industry experience, working on various construction projects in collaboration with leading construction companies. This hands-on experience has provided him with valuable insights into the practical aspects of construction management, enabling him to bridge the gap between theory and real-world application.

Smith's contributions to the field of construction management extend beyond his academic pursuits. He is a prolific author, having published numerous articles in reputable journals and presented his research findings at international conferences. His work has explored diverse topics such as construction productivity, health and safety, and sustainable construction practices. Smith's publications have made a significant impact on the industry, helping professionals and scholars alike to enhance their understanding and improve their practices in construction management.

As a highly regarded academic and mentor, Smith has guided and inspired countless students pursuing careers in construction management. His dedication to education and his ability to convey complex concepts in a clear and accessible manner have earned him the respect and admiration of his students.

Throughout his career, Smith has demonstrated a commitment to continuous learning and professional development. In his spare time, Smith enjoys traveling to construction sites, exploring new construction technologies, and engaging in conversations with industry professionals to stay at the forefront of industry trends.

With his vast knowledge, extensive experience, and dedication to the field, Steven Smith brings a wealth of expertise to his role as an author, ensuring that his book provides valuable insights and practical guidance to construction professionals, academics, and students alike.

BOOKS BY THIS AUTHOR

Construction Health And Safety Fundamentals

In the fast-paced world of construction, ensuring the health and safety of workers is paramount. This book is an indispensable guide that lays the foundation for creating a culture of safety and excellence in the construction industry.

Authored by a seasoned expert in construction health and safety, the book offers a comprehensive exploration of the fundamental principles, strategies, and best practices that underpin effective safety management. It covers a wide range of topics, including safe operation of heavy machinery, crane and hoist safety, rigging and lifting operations, power tool safety, electrical safety, fire safety and emergency response, working at heights and fall protection, confined space entry and rescue, and much more.

Designed as a practical resource, each chapter provides clear explanations of key concepts and actionable insights. One of the distinguishing features of this book is its comprehensive approach to emerging trends and technologies in construction health and safety. It explores innovative solutions such as wearable technologies, virtual reality training, and predictive analytics, empowering readers to stay ahead of the curve and leverage cutting-edge tools for enhanced safety outcomes.

Whether you are a construction professional, safety manager, project supervisor, or worker looking to enhance your knowledge and skills in construction health and safety, this book is your go-to resource. Don't compromise on safety! Together, we can build a safer and more prosperous future for the construction industry.

Construction Management Blueprint: A Comprehensive Guide To Successful Project Delivery

Are you ready to take your construction management skills to the next level? Look no further! This book is the ultimate guidebook that will equip you with the knowledge and tools you need to excel in the complex world of construction projects.

This comprehensive book covers every aspect of construction management, from project initiation to post-construction evaluation. Written by industry experts with decades of experience, it provides a wealth of practical insights, real-life case studies, and invaluable tips for success.

Inside these pages, you'll find:
A step-by-step guide to project initiation and feasibility studies, helping you identify objectives, assess market demand, and engage stakeholders effectively.

In-depth coverage of project planning and design, including goal setting, scope definition, work breakdown structures, and sustainable design principles.
Extensive discussions on cost estimation techniques, budgeting, resource allocation, value engineering, and contract pricing and negotiation strategies.
Detailed insights into construction scheduling, resource procurement, site layout and logistics, and risk management, ensuring smooth project execution.

Thorough examinations of quality assurance and control, materials testing and inspection, occupational health and safety practices, and risk mitigation strategies.

Expert guidance on commissioning and handover, facility documentation, owner training, and maintenance for long-term sustainability.

Essential information on post-construction evaluation, continuous improvement, professional development, and knowledge management in construction management.

The book is not just a theoretical guide; it's a practical companion for construction professionals, project managers, and students looking to enhance their skills and achieve outstanding results. Its clear and concise language, combined with visually engaging diagrams and insightful case studies, makes it an enjoyable and accessible read for both industry veterans and newcomers.

Whether you're involved in residential, commercial, or infrastructure projects, the book will empower you to navigate the complexities of the construction industry with confidence and achieve excellence in every aspect of your work.

Don't miss out on this opportunity to unlock your full potential as a construction management professional. Get your copy of the book today and embark on a journey toward unparalleled success in the dynamic world of construction projects.

The Dictionary Of Construction Terminologies: A Compendium Of Knowledge For Students, Academics, Practitioners, And House Owners

The dictionary of construction terminologies book is a comprehensive reference guide that provides definitions and explanations of the technical language and jargon used in the construction industry. It is an invaluable resource for professionals working in construction, as well as for students learning about the industry or for individuals looking to understand construction-related concepts better.

The book features a wide range of entries that cover various aspects of construction, including architecture, engineering, materials, equipment, and techniques. The book also provides clear and concise definitions of technical terms, written in easy-to-understand language. Terminologies are presented alphabetically to help readers find the descriptions they need quickly and easily. Whether you are a professional working in the field or interested in construction, this book is an essential tool to help you navigate the complex world of construction terminology with confidence and clarity.

Are you a student of construction, a house owner, an academic in the construction industry, or a practitioner that desires to acquire more knowledge about construction terms? If your answer to the preceding question is affirmative, this book may be one of the best investments you will ever make.

Construction Management Fundamentals: A Handbook For Construction Students, Academics, And Practitioners

Are you seeking a comprehensive guide to mastering the fundamentals of construction management? Look no further than this book! This handbook offers a clear and concise overview of the principles, practices, and techniques of construction management. It covers the concepts essential for a successful career in construction management, including project planning, scheduling, and construction contracts. The book is designed to be an invaluable resource for construction students, academics, and practitioners alike.

Whether you're a seasoned professional looking to refresh your skills or a student just starting out in the field, this handbook is an indispensable resource that will help you excel in your career.

Get your copy today and take your construction management skills to the next level!

Building Maintenance Guidelines: A Complete Manual

This book is an essential resource for anyone responsible for maintaining and preserving the integrity of a building. It covers several aspects of building maintenance, from electrical systems and HVAC systems to roofing, plumbing, and structural components. It provides clear, step-by-step instructions on how to perform routine maintenance tasks. It also includes information on how to identify potential problems, such as water damage, mold growth, and insect infestations, and provides guidance on how to address these issues. In addition to its practical information, the book also includes important information on energy efficiency and sustainability.

With its clear, easy-to-follow language, the book is an invaluable resource for anyone looking to keep their building in optimal condition.

Aerogel As A Sustainable Construction Material: Towards Net Zero Carbon Emissions

This book is a comprehensive guide for anyone interested in learning about a groundbreaking material (Aerogel) and its potential to revolutionize sustainable building practices. The book provides a detailed overview of aerogel, its unique properties, and its applications in the construction industry. It explains how aerogel can be used to promote energy efficiency and reduce carbon emissions, making it an essential tool for achieving net-zero carbon footprints in construction projects.

Written by an expert in the field, this book covers a range of topics, including the science behind aerogel, its production and manufacturing, and its use in different building applications. It also includes case studies and real-world examples that showcase the practical applications of aerogel in sustainable construction.

The book is recommended for construction professionals, researchers, students, and people that are simply interested in learning about the latest developments in sustainable building practices. With its clear and accessible writing style, engaging content, and practical insights, the book is sure to inspire and inform readers about the potential of aerogel as a game-changing material in the pursuit of sustainable construction practices.

www.ingramcontent.com/pod-product-compliance
Lightning Source LLC
Chambersburg PA
CBHW070424240526
45472CB00020B/1196